NIST Technical Note 1626

ALGORITHMS FOR GENERATING LARGE SETS OF SYNTHETIC DIRECTIONAL WIND SPEED DATA FOR HURRICANE, THUNDERSTORM, AND SYNOPTIC WINDS

Mircea Grigoriu
Building and Fire Research Laboratory
National Institute of Standard and Technology
Gaithersburg, MD 20899-8611, and

School of Civil and Environmental Engineering
Cornell University
Ithaca, NY 14850

February 2009

TABLE OF CONTENTS

==

ACKNOWLEDGMENTS

==

This work was performed under the terms of the Intergovernmental Personnel Act. The author wishes to acknowledge valuable contributions by Dr. J. A. Main of the Building and Fire Research Laboratory (BFRL). Useful exchanges with Dr. E. Simiu of BFRL, who initiated this study, are also acknowledged with thanks.

LIST OF ACRONYMS AND ABBREVIATIONS

===

MRI   Mean Recurrence Interval

LIST OF FIGURES

==

Disclaimers

Certain commercial entities, equipment, or materials may be identified in this document in order to describe an experimental procedure or concept adequately. Such identification os not intended to imply recommendation or endorsement by the National Institute of Standard and Technology, nor is it intended to imply that the entities, materials, or equipment are necessarily the best available for the purpose.

The policy of the NIST is to use the International System of Units in its technical communications. However, in North America in the construction and building materials industry, certain non-SI units are so widely used instead of SI units that it is more practical and less confusing to include measurement values for customary units only.

ABSTRACT

===

Probabilistic models are developed for directional wind speeds in hurricane, thunderstorm, and synoptic wind storms. The models, calibrated to data, are used to generate synthetic directional wind speeds over periods of arbitrary length and assess the uncertainty in the resulting extreme wind speeds. The generation uses MATLAB functions, called **dir_hurricane**, **bootstrap_par**, and **bootstrap_par_ts**, that are available on www.nist.gov/wind. The synthetic data generated by the models and MATLAB functions developed in this study provide a rational tool for constructing synthetic directional wind speed data that are statistically consistent with existing wind records. The developments in this study are needed because large sets of synthetic data are required to assess the performance of tall buildings and other structural systems under directional wind speeds.

Chapter 1
INTRODUCTION

The assumption that wind effects are proportional to the square of the extreme wind speeds, regardless of their direction, is used in building codes for the design of most rigid structures. This assumption allows the estimation of wind effects with large mean recurrence intervals (MRIs) from probabilistic models of extreme wind speeds calibrated to relatively small observed data sets. However, for structures sensitive to wind directionality effects, including special rigid structures and structures with significant dynamic response, the estimation of wind effects with large MRIs requires the use of time series of directional wind speeds with length exceeding the length of the MRI of interest. To overcome this difficulty, it is proposed to construct synthetic directional wind speed records that (1) have any specified length and (2) are statistically consistent with available records. The construction of synthetic wind records involves two steps. First, probabilistic models are developed for directional wind speeds, and these models are calibrated to data. Second, the calibrated models are used to generate synthetic directional wind speed records.

Data available for calibrating the proposed probabilistic models for directional wind speeds are stored in matrices with n rows and d columns. The elements in the rows of these matrices are wind speeds recorded in d directions during distinct wind events. The calibration of the probabilistic models for directional wind speeds poses notable difficulties since in practice many of the entries of the wind data matrices can be zero, so that the number of non-zero wind speeds can be much smaller than the number of wind events. This situation is particularly severe for thunderstorm and synoptic winds since both the number n of wind events and the number of non-zero wind speeds are relatively small. Accordingly, the construction and calibration of probabilistic models for thunderstorm and synoptic winds must in many if not most cases be based, in addition to data, on heuristic considerations.

The probabilistic model for hurricane wind speeds is sufficiently general to describe directional wind speeds in thunderstorms and synoptic storms. Nevertheless, if the number of observations available for these storms is small, the hurricane model may not be useful for modeling their wind speeds. For hurricane winds, the probabilistic model is based on translation vectors, discussed later in this report, and accounts in an approximate but conservative manner for the correlation between directional wind speeds. For thunderstorm and synoptic winds, the correlation between directional wind speeds is not modeled because the available records are insufficient to estimate it reliably. The arrival rate of hurricane, thunderstorm, and synoptic events is modeled by a homogeneous Poisson processes with mean rate inferred from observations. The proposed probabilistic models calibrated to data are used to develop MATLAB codes that (1) generate directional wind speeds for hurricane, thunderstorm, and synoptic winds over specified periods and (2) assess the uncertainty in the generated wind speeds using bootstrapping methods.

The report is organized as follows. Basic considerations on the probabilistic models for directional wind speeds for hurricane, thunderstorm, and synoptic winds are presented in Chapter 2. This chapter also explores effects of correlation between directional wind speeds. Chapter 3 defines wind speeds with specified MRIs, presents calculations based on

1

nominal structural demands with various MRIs, and examines the sector-by-sector method for calculating wind speeds with given MRIs. Probabilistic wind speeds for hurricane winds are defined and calibrated in Chapter 4. This chapter also outlines the steps of the Monte Carlo algorithm used to generate directional wind speeds in hurricanes. Chapter 5 defines and calibrates probabilistic models for directional wind speeds in thunderstorm and synoptic winds, and presents a Monte Carlo algorithm for generating directional wind speeds for thunderstorm and synoptic events. Chapter 6 demonstrates the application of the MATLAB functions for hurricane winds, including the construction of replicates for predicted wind speeds with various MRIs obtained by bootstrapping. Concluding remarks are presented in Chapter 7. Two appendices provide detailed information on the use of the MATLAB functions for generating directional wind speeds and for uncertainty assessment. These computer codes, named **dir_hurricane**, **bootstrap_par**, and **bootstrap_par_ts**, are available on www.nist.gov/wind.

Chapter 2
WIND SPEED MODEL
===

Let $T_1 < T_2 < \cdots$ denote the sequence of random arrival times of hurricane or other extreme wind storms at a site. It is assumed that the random times $\{T_k, k = 1, 2, \ldots\}$ are the jump times of an homogeneous Poisson process $\{N(t), t \geq 0\}$ of intensity $\lambda > 0$, so that the average number of extreme wind events during a time $t > 0$ is equal to the expectation $E[N(t)] = \lambda t$ of $N(t)$. The intensity of N can be estimated from data. For example, suppose we have observed 30 extreme wind events at a site during a 10 year period. An estimate of λ is $\hat{\lambda} = 30/10 = 3$ events/year. The time arrival model of extreme wind events can easily be generalized to account for the possible seasonality of extreme wind events by letting λ to be time dependent.

Let $\boldsymbol{V}^{(k)}$ be a d-dimensional wind speed vector whose coordinates $\{V_{k,i}, i = 1, \ldots, d\}$ are wind speeds recorded in d directions during the kth extreme wind event, that is, the wind speed vector at time T_k. It is assumed that the wind speed vectors $\{\boldsymbol{V}^{(k)}, k = 1, 2, \ldots\}$ are independent copies of a d-dimensional random vector $\boldsymbol{V}$ with joint distribution F. In view of the above considerations and assumptions, the proposed model for extreme wind events is completely characterized by (1) the intensity λ of the Poisson model N and (2) the distribution of $\boldsymbol{V}$. As previously mentioned, the intensity λ of the Poisson process N can be estimated from data by elementary calculations. However, the selection of the joint distribution F of $\boldsymbol{V}$ is much more difficult because there are no general models for arbitrary non-Gaussian joint distributions and the available data is insufficient for constructing reliable empirical distributions for $\boldsymbol{V}$.

Two different models are used for the distribution of $\boldsymbol{V}$. For hurricanes, we assume that $\boldsymbol{V}$ is a translation vector, that is, a nonlinear transformation of a d-dimensional Gaussian vector. Translation vectors account in an approximate manner for the correlation between directional wind speeds. Data on hurricanes available, for example, in the NIST database (www.nist.gov/wind), are sufficient to calibrate this model of $\boldsymbol{V}$, which can also be used for thunderstorm and synoptic winds if sufficient data were available for these types of storms. On the other hand, translation vectors cannot be used to describe thunderstorm and synoptic winds if, as is the case for the data available for this study (that is, Automated Surface Observing System (ASOS) data for Newark airport, NJ, LaGuardia airport, NY, and Kennedy airport, NY), the number of observations is insufficient. For these types of storms and data, we assume therefore that directional wind speeds are independent random variables.

2.1 DIRECTIONAL WIND SPEEDS

Suppose the coordinates $\{V_i\}$ of the wind speed vector $\boldsymbol{V}$ are defined by

$$V_i = F_i^{-1}\big(\Phi(G_i)\big), \quad i = 1, \ldots, d, \tag{1}$$

where $\{F_i, \ i = 1, \ldots, d\}$ denote arbitrary distributions, Φ denotes the distribution of the standard Gaussian variable with mean 0 and variance 1, and $\{G_i, i = 1, \ldots, d\}$ are correlated standard Gaussian variables with covariance matrix $\boldsymbol{\rho} = \{\rho_{ij} = E[G_i G_j], \ i, j = 1, \ldots, d\}$. We note that (1) $\{F_i\}$ are the distributions of the coordinates $\{V_i\}$ of $\boldsymbol{V}$, (2) Eq. 1 establishes a one-to-one correspondence between the wind speed vector $\boldsymbol{V}$ and the Gaussian vector $\boldsymbol{G} = (G_1, \ldots, G_d)$, (3) the correlation coefficients between (G_i, G_j) and (V_i, V_j) are similar for positive values of ρ_{ij}, and (4) $\boldsymbol{V}$, referred to as a translation vector, is completely defined by the marginal distributions $\{F_i, \ i = 1, \ldots, d\}$ and the covariance matrix $\boldsymbol{\rho}$ of $\boldsymbol{G}$ ([1], Section 3.1.1).

Under the translation model in Eq. 1, the resulting model for directional extreme winds is completely defined by the intensity λ of the Poisson process N, the covariance matrix $\boldsymbol{\rho}$ of the Gaussian image $\boldsymbol{G}$ of $\boldsymbol{V}$, and the marginal distributions $\{F_i\}$ of $\boldsymbol{V}$. These parameters can be estimated from data and used subsequently in a Monte Carlo algorithm to generate virtual wind time series of any length. As previously stated, we use the representation in Eq. 1 for hurricane winds.

For thunderstorm and synoptic winds, we set $\boldsymbol{\rho}$ equal to the identity matrix, so that the coordinates of $\boldsymbol{G}$ are independent standard Gaussian variables. Accordingly, the random variables $\{V_i\}$ defined by Eq. 1 are independent and follow the distributions $\{F_i\}$.

The following section examines dependence of wind speed maxima over wind directions and elementary structural demands on the correlation between the coordinates $\{V_i\}$ of the wind speed vector $\boldsymbol{V}$. These considerations are relevant only for models of hurricane winds.

2.2   CORRELATED DIRECTIONAL WIND SPEEDS

Let $\boldsymbol{G}'$ and $\boldsymbol{G}''$ be d-dimensional standard Gaussian vectors with coordinates of mean 0, variance 1, and covariance matrices $\boldsymbol{\rho}' = \{\rho_{ij}' = E[G_i' G_j']\}$ and $\boldsymbol{\rho}'' = \{\rho_{ij}'' = E[G_i'' G_j'']\}$ such that $\rho_{ij}' \leq \rho_{ij}''$, $i, j = 1, \ldots, d$. Then

$$P\left(\cap_{i=1}^{d} \{G_i' \leq \xi_i\}\right) \leq P\left(\cap_{i=1}^{d} \{G_i'' \leq \xi_i\}\right) \tag{2}$$

by a corollary to the normal comparison lemma ([5], Corollary 4.2.3). This inequality holds for the special case in which $\boldsymbol{G}'$ has independent coordinates and $\boldsymbol{G}''$ has positively correlated coordinates, that is, $\rho_{ij}' = 0$, $i \neq j$, and $\rho_{ij}'' \geq 0$ for $i, j = 1, \ldots, d$.

The inequality in Eq. 2 extends directly to the wind speed vector $\boldsymbol{X}$ defined by Eq. 1, since the events $\{V_i \leq x_i\}$ and $\{G_i \leq \Phi(F_i(x_i))\}$ have the same probability, $\Phi(F_i(x_i))$, $i = 1, \ldots, d$, are monotonically increasing functions, and the correlation coefficient between two distinct coordinates (V_i, V_j) of $\boldsymbol{V}$ is an increasing function of the correlation coefficient ρ_{ij} between the coordinates (G_i, G_j) of $\boldsymbol{G}$ ([1], Section 3.1.1).

Let $\boldsymbol{V}' = \{V_i'\}$ and $\boldsymbol{V}'' = \{V_i''\}$ be wind speed vectors defined by Eq. 1 with $\boldsymbol{G}'$ and $\boldsymbol{G}''$ in Eq. 2 in place of $\boldsymbol{G}$. Recall that $\boldsymbol{G}'$ and $\boldsymbol{G}''$ are standard Gaussian vectors with covariance matrices such that $\rho_{ij}' \leq \rho_{ij}''$, $i, j = 1, \ldots, d$. Then

$$P\left(\cap_{i=1}^{d} \{X_i' \leq x_i\}\right) \leq P\left(\cap_{i=1}^{d} \{X_i'' \leq x_i\}\right), \tag{3}$$

where $x_i = F_i^{-1}(\Phi(\xi_i))$, $i = 1, \ldots, d$. Let $F_{\max}'$ and $F_{\max}''$ be the distributions of $V_{\max}' = \max_{1 \leq i \leq d}\{V_i'\}$ and $V_{\max}'' = \max_{1 \leq i \leq d}\{V_i''\}$, respectively. The MRIs R' and R'' for a wind

4

speed v irrespective of direction under the models $\boldsymbol{V}'$ and $\boldsymbol{V}''$, respectively, of a wind speed vector $\boldsymbol{V}$ satisfy the inequality

$$R'(v) = \frac{1}{1 - F'_{\max}(v)} \leq \frac{1}{1 - F''_{\max}(v)} = R''(v), \tag{4}$$

showing that the MRI of a wind speed v is shorter under the model with weaker correlation between directional wind speeds. In other words, if the actual correlation of a wind speed vector $\boldsymbol{V}$ is underestimated, the resulting wind speeds are conservative since they have shorter MRIs than those corresponding to the correct correlation between directional wind speeds.

Consider the special case in which $\boldsymbol{G}'$, and, therefore, $\boldsymbol{V}'$ has independent coordinates, that is, $\rho'_{ij} = 0$, $i \neq j$, and $\boldsymbol{V}''$ has positively correlated coordinates, that is, $\rho''_{ij} \geq 0$. Then, a relationship as in Eq. 4 holds with R', so that the assumption of independence between directional wind speeds is conservative if the actual correlation between directional wind speeds are positive and the wind speed can be modeled by a translation vector.

For illustration, suppose the coordinates of the Gaussian image $\boldsymbol{G}$ of $\boldsymbol{V}$ in Eq. 1 are equally and positively correlated, that is $E[G_i G_j] = \rho \geq 0$, $i, j = 1, \ldots, d$, $i \neq j$. The vector $\boldsymbol{G}$ can be defined by

$$G_i = \sqrt{\rho}\, W + \sqrt{1 - \rho}\, W_i, \quad i = 1, \ldots, d, \tag{5}$$

where W and W_i, i=1,...,d, are independent standard Gaussian variables with mean 0 and variance 1. That the variables G_i have the stated properties follow by direct calculations. The coordinates of the image $\boldsymbol{V}$ of $\boldsymbol{G}$ are also equally correlated ([1], Section 3.1.1).

We present two examples illustrating the dependence of $F_{\max}$ and wind speeds with various MRIs on the correlation coefficient ρ of $\boldsymbol{G}$, assumed to be positive.

Example 1. Records of yearly wind speed maxima observed in $d = 8$ directions at Tucson, Arizona, from 1959 to 1970 yield means and standard deviations of wind speeds in these directions of $\{\mu_i, i = 1, \ldots, 8\} = \{30.40, 35.93, 42.13, 44.37, 39.90, 37.97, 41.40, 35.73\}$ in mph and $\{\sigma_i, i = 1 \ldots, 8\} = \{7.85, 11.34, 7.91, 9.26, 7.34, 4.31, 9.25, 6.74\}$ in mph. Let $\boldsymbol{V}$, $d = 8$, denote these directional wind speeds assumed to follow the Extreme Type I distributions

$$F_i(x) = \exp\left(-\exp(-\alpha_i (x - u_i))\right), \quad x \in \mathbb{R}, \tag{6}$$

where $\mu_i = u_i + 0.577216/\alpha_i$ and $\sigma = \pi/(\sqrt{6}\, \alpha_i)$. We are interested in properties of the largest wind speeds $V_{\max}$ irrespective of direction, that is, the largest coordinate of $\boldsymbol{V}$. Figure 1 shows wind speeds $V_{\max}$ in Tucson with MRIs up to 10,000 years under the assumption that the coordinates of $\boldsymbol{V}$ are independent and equally correlated with correlation coefficients $\rho = 0.7$ and $\rho = 0.999$. Differences between wind speeds corresponding to independent and correlated coordinates of $\boldsymbol{V}$ are negligible for $\rho = 0.7$. These differences become significant for wind speed vectors with strongly correlated coordinates, for example, directional wind speeds corresponding to $\rho = 0.999$.

We note that (1) the independence assumption yields conservative approximations in agreement with our previous theoretical arguments and (2) $V_{\max}$ provides a crude measure of demand at a cross section in a structure if characterized by the same internal forces irrespective of wind direction.

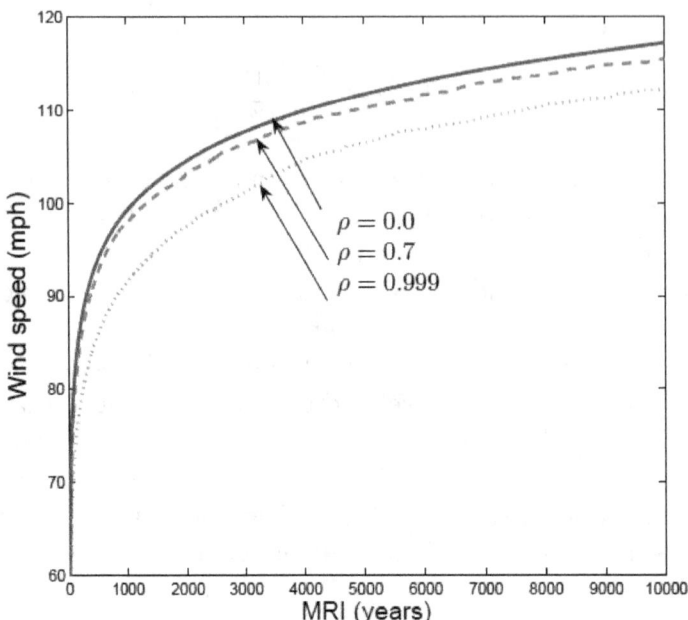

Figure 1: Wind speeds in mph with MRIs up to 10,000 year at Tucson, Arizona

Example 2. Consider the wind environment in the previous examples and a structural demand assumed to be dominated by wind speeds from a single direction, say direction 1, so that $P(V_1 > x) \gg P(V_i > x)$ or, equivalently, $F_1(x) \ll F_i(x) \simeq 1$ for $i = 2, \ldots, 8$. If we view wind speed as a measure of structural demand, the distribution of $V_{\max}$ is

$$P(V_{\max} \le x) = \prod_{i=1}^{8} F_i(x) \simeq F_1(x), \tag{7}$$

and

$$
\begin{aligned}
P(V_{\max} \le x) &= P\big(\cap_{i=1}^{8} \{V_i \le x\}\big) = P\big(V_1 \le x, \alpha_2 V_1 + \beta_2 \le x, \ldots, \alpha_8 V_1 + \beta_8 \le x\big) \\
&= P\big(V_1 \le x, V_1 \le (x - \beta_2)/\alpha_2, \ldots, V_1 \le (x - \beta_8)/\alpha_8\big) \simeq P\big(V_1 \le x\big) = F_1(x) \quad (8)
\end{aligned}
$$

under the assumptions that the coordinates of $\boldsymbol{V}$ are independent and perfectly correlated, respectively, where $V_i = \alpha_i V_1 + \beta_i$, $i = 2, \ldots, 8$, for perfectly correlated directional wind speeds and $(\alpha_i > 0, \beta_i)$ are constants such that $x \ll (x - \beta_i)/\alpha_i$. In this case, the dependence between directional wind speeds has a negligible effect on nominal structural demand.

Chapter 3
MEAN RECURRENCE INTERVALS FOR
WIND SPEEDS AND NOMINAL DEMANDS

Let $X_1, X_2, \ldots$ be a sequence of independent identically distributed real-valued random variables with distribution F_x and N a homogeneous Poisson process of intensity λ. Consider the random series taking the values $X_1, X_2, \ldots$ at the jump times $0 < T_1 < T_2 < \cdots$ of N. The average time between the jump times of N is $1/\lambda$.

Suppose we retain from the above time series only those values exceeding an arbitrary threshold a. The random times $0 < T_1' < T_2' < \cdots$ of the non-zero values of the resulting series define a new homogeneous Poisson process of intensity $\lambda(a) = \lambda\left(1 - F_x(a)\right)$. If λ is measured in jumps/year, then $1/\lambda(a)$ gives the average number of years between consecutive values of the original series exceeding a. The solution a_R of

$$R = \frac{1}{\lambda\left(1 - F_x(a_R)\right)} \tag{9}$$

is a level exceeded on average once every R years, and is referred to as threshold or level with MRI R.

3.1   WIND SPEEDS AND STRUCTURAL DEMANDS

The joint distribution of $\boldsymbol{V}$ is

$$F(\boldsymbol{x}) = P\left(\cap_{i=1}^{d} \{X_i \leq x_i\}\right) = P\left(\cap_{i=1}^{d} \{G_i \leq \xi_i\}\right), \tag{10}$$

where $\xi_i = F_i^{-1}\left(\Phi(x_i)\right)$ denotes the image of x_i, $i = 1, \ldots, d$, in the Gaussian space (Eq. 1). Since the events $\{V_{\max} = \max_{1 \leq i \leq d}\{V_i\} \leq x\}$ and $\cap_{i=1}^{d}\{V_i \leq x\}$ have the same probability, the distribution $F_{\max}$ of the largest wind speed $V_{\max}$ irrespective of direction is

$$F_{\max}(x) = P\left(V_{\max} \leq x\right) = P\left(\cap_{i=1}^{d} \{V_i \leq x\}\right) = P\left(\cap_{i=1}^{d} \{G_i \leq \xi\}\right), \tag{11}$$

where $\xi = F_i^{-1}\left(\Phi(x)\right)$.

Distributions of the type defined by Eq. 11 can be used to calculate wind speeds with specified MRIs. For example, let $\boldsymbol{V}^{(1)}, \boldsymbol{V}^{(2)}, \ldots$ be a wind speed series consisting of independent copies of $\boldsymbol{V}$ arriving at random times $0 < T_1 < T_2 < \cdots$ defined by the jumps of a homogeneous Poisson process of intensity λ jumps/year. Suppose we are interested in the sequence $V_{1,\max}, V_{2,\max}, \ldots$ of extreme wind speeds irrespective of direction. The wind speed v_R of this series of MRI R years is the solution of (Eq. 9)

$$R = \frac{1}{\lambda\left(1 - F_{\max}(v_R)\right)}. \tag{12}$$

Similar results hold for structural demands under wind loads. Let $D_s^{(1)}, D_s^{(2)}, \ldots$ be a real-valued series of demands at a critical point s of a structure corresponding to a wind

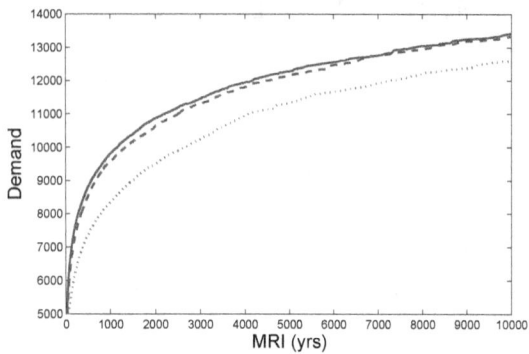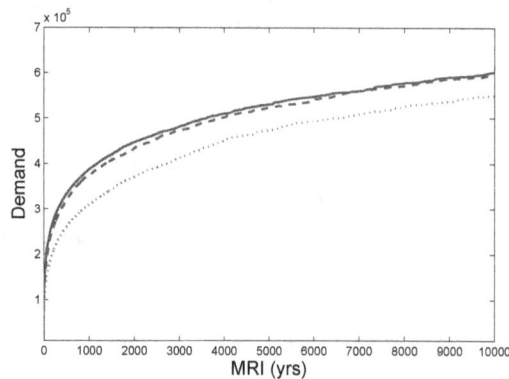

Figure 2: Nominal structural demands for $q = 2$ (left panel) and $q = 2.8$ (right panel) with MRIs up to 10,000 year for Tucson, Arizona (non-dimensional)

speed series $\boldsymbol{V}^{(1)}, \boldsymbol{V}^{(2)}, \dots$ with time step equal to the jump times of a homogeneous Poisson process of intensity λ events/year. The construction of the demand series involves two steps. First, design interaction formulas are used to calculate demands $\{D_{s,i}^{(k)}, i = 1, \dots, d\}$, corresponding to the coordinates $\{V_i^{(k)}, i = 1, \dots, d\}$ a wind speed vector $\boldsymbol{V}^{(k)}$. Second, the demand sequence of interest in design is the series $D_s^{(k)} = \max_{1 \le i \le d}\{D_{s,i}^{(k)}\}$. The demand $d_{R,s}$ of MRI R years at point s is the solution of (Eq. 9)

$$R = \frac{1}{\lambda\left(1 - F_{D_s}(d_{R,s})\right)} , \tag{13}$$

where F_{D_s} denotes the distribution of D_s.

Example 3. Nominal structural demand for rigid and flexible structures is proportional to wind speed at power 2 and, say, 2.8, respectively ([7], p. 177). Accordingly, structural demands at a point s in a structure caused by directional wind speeds $\{V_i\}$ are assumed to have the order of magnitude $D_{s,i} = V_i^q$, where $q = 2$ and $q = 2.8$ for rigid and flexible structures, respectively.

Figure 2 shows quasi-static and dynamic responses with MRIs up to 10,000 years generated by the directional wind speed in Example 1, that is, equally correlated Extreme Type 1 directional wind speeds calibrated to data in Tucson, Arizona. The solid, dashed, and dotted lines correspond to correlation coefficients $\rho = 0.0$, 0.7, and 0.999. As for extreme winds in Fig. 1, demands for both rigid and flexible structures are not sensitive to correlation, unless the correlation is very strong, that is, the independence model is conservative.

3.2   SECTOR-BY-SECTOR METHOD

The sector-by-sector method defines structural demands with an MRI R as the largest of the directional structural demands with this return period. Let $\{d_{R,si}, i = 1, \ldots, d\}$ denote structural demands with MRI R at point s in a structure corresponding to directional winds. These demands are solutions of

$$R = \frac{1}{\lambda \left(1 - F_{D_{s,i}}(d_{R,si})\right)}, \tag{14}$$

where $F_{D_{s,i}}$ is the distribution of demand at location s under wind speed V_i from direction i. It is assumed that structural demand with MRI R is $d^*_{R,s} = \max_{1 \leq i \leq d} d_{R,si}$.

Since

$$F_{D_s}(\xi) = P\left(\cap_{j=1}^d \{D_{s,j} \leq \xi\}\right) \leq P\left(\cap_{i=1}^d \{D_{s,i} \leq \xi\}\right) = F_{D_{s,i}}(\xi) \tag{15}$$

and

$$1 - 1/R = F_{D_{s,i}}(d_{R,si}) = F_{D_s}(d_{R,s}) \leq F_{D_{s,i}}(d_{R,s}), \tag{16}$$

we have $d_{R,s} \leq d_{R,si}$, $i = 1, \ldots, d$, by the monotonicity of distribution functions, implying $d_{R,s} \leq d^*_{R,s} = \max_{1 \leq i \leq d} d_{R,si}$. Hence, R-year demands defined by the sector-by-sector method over-predict actual R-year demands, showing that the method is unconservative.

Chapter 4
WIND MODEL FOR HURRICANE WINDS

==

The proposed model for directional extreme wind speeds is completely specified by a Poisson process N of intensity λ describing the time arrival of extreme wind events, and a translation random vector $\boldsymbol{V}$ defining directional wind speeds in individual extreme wind events. The defining parameters of the model are the intensity λ of N, the distributions $\{F_i\}$ of the coordinates $\{V_i\}$, $i = 1, \ldots, d$, of $\boldsymbol{V}$, and the covariance matrix $\{\rho_{ij}\}$, $i, j = 1, \ldots, d$, of the Gaussian image $\boldsymbol{G}$ of $\boldsymbol{V}$.

Our objectives are to (1) calibrate the wind model to data, (2) develop a Monte Carlo algorithm for generating synthetic extreme wind speed records of any length that are consistent with the available information, and (3) assess the sensitivity to statistical uncertainty in wind speeds and structural demands with large MRIs.

4.1   MODEL CALIBRATION

Suppose that wind data consist of records of extreme directional wind speeds in d directions recorded during n distinct events occurring over an n_y year period. We use these data to estimate the defining parameters of the wind model, that is, the intensity λ of N and the probability law of $\boldsymbol{V}$.

The average number λ of extreme wind speed events per year can be estimated by the ratio $\hat{\lambda} = n/n_y$. This estimator of λ is unbiased and has small uncertainty for the available data since n is of the order 1,000. It is assumed that $\hat{\lambda} = n/n_y$ is the actual value of λ.

The calibration of the probability law of $\boldsymbol{V}$ is less simple and consists of two steps. First, we estimate the the distributions $\{F_i\}$ of the directional wind speeds $\{V_i\}$. Second, we estimate the covariance matrix $\boldsymbol{\rho}$ of the Gaussian image $\boldsymbol{G}$ of $\boldsymbol{V}$. We proceed with the first step. Since various directional wind speed data are zero, we model the marginal distributions $\{F_i\}$ of the directional wind speeds $\{V_i\}$ by

$$F_i(x) = q_i\, 1(x \geq 0) + (1 - q_i)\, \tilde{F}_i(x), \quad i = 1, \ldots, d, \tag{17}$$

where $1(A)$ denotes the indicator of set A, a function equal to 1 and 0 if A is true and false, respectively, $q_i = P(V_i = 0)$ is the probability that V_i is 0 in an arbitrary extreme wind event, and $\tilde{F}_i$ is a proper distribution characterizing the non-zero values of V_i. The functional forms of the distributions $\tilde{F}_i$ are assumed known. We need to estimate both the probabilities $q_i = P(V_i = 0)$ and the parameters of the distributions $\tilde{F}_i$, $i = 1, \ldots, d$.

The probability q_i can be estimated by the ratio $\hat{q}_i = n_i/n$, where n_i is the number of observed zero wind speeds in direction $i = 1, \ldots, d$. This estimator of q_i is unbiased, that is, its mean is q_i, with variance $q_i\,(1 - q_i)/n$ and coefficient of variation $\sqrt{(1 - q_i)/q_i}/\sqrt{n}$. For the data set under consideration, $n = 1,000$ and the probabilities q_i are usually in the range 0.4 to 0.9 so that its estimators are typically accurate. For example, for $n = 1,000$ and $q_i = 0.5$, the coefficient of variation of the estimator of q_i is $1/\sqrt{n} = 0.0316$. The relatively small uncertainty in the estimators of q_i suggest to approximate q_i by their estimates $\hat{q}_i$.

11

We now estimate the parameters of the distributions $\{\tilde{F}_i\}$ assumed to have known functional forms by using only the non-zero readings of directional wind speeds $\{V_i\}$. For example, suppose $\{\tilde{F}_i\}$ are reverse Weibull distributions with parameters (α_i, η_i, c_i), and let Y be a Weibull random variable with parameters $\alpha > 0$, $\xi \in \mathbb{R}$, and $c > 0$, distribution

$$F(y) = \begin{cases} 1 - \exp\left[-\left(\frac{y-\xi}{\alpha}\right)^c\right], & y > \xi \\ 0 & y \le \xi, \end{cases} \tag{18}$$

and density

$$f(y) = \frac{c}{\alpha}\left(\frac{y-\xi}{\alpha}\right)^{c-1}\exp\left[-\left(\frac{y-\xi}{\alpha}\right)^c\right], \quad y > \xi. \tag{19}$$

Values of distribution $F(y)$ and solutions of $F(y_p) = p \in [0,1]$ can be calculated by, for example, the MATLAB functions $F(y) = \mathrm{cdf}(\text{'wbl'}, y-\xi, \alpha, c)$ and $y_p - \xi = \mathrm{icdf}(\text{'wbl'}, p, \alpha, c)$. The random variable $X \overset{d}{=} -Y$ has the distribution

$$F(x) = P(X \le x) = P(Y > -x) = \exp\left[-\left(\frac{\eta-x}{\alpha}\right)^c\right], \quad x \le \eta = -\xi, \tag{20}$$

and is called reverse Weibull variable. The expression of F in Eq. 20 coincides with that in [8] (Equation A1.65a) depending on the parameters (σ, μ, γ) denoted here by (α, η, c). Moments of any order of Y can be obtained from moments $E[\tilde{Y}^q] = \Gamma(1+q/c)$ of the scaled random variable $\tilde{Y}$ defined by $Y = \xi + \alpha\tilde{Y}$ ([3], Chapter 20). For example, the mean μ_y, variance σ_y^2, and skewness $\gamma_{y,3}$ of Y are

$$\mu_y = \xi + \alpha\,\Gamma(1+1/c)$$
$$\sigma_y^2 = \alpha^2\left(\Gamma(1+2/c) - \Gamma(1+1/c)^2\right)$$
$$\gamma_{y,3} = \frac{\Gamma(1+3/c) - 3\,\Gamma(1+1/c)\,\Gamma(1+2/c) + 2\,\Gamma(1+1/c)^3}{\left(\Gamma(1+2/c) - \Gamma(1+1/c)^2\right)^{3/2}}. \tag{21}$$

Let $v_{i,1}, v_{i,2}, \ldots, v_{i,n-n_i}$ be non-zero wind speed data in a direction $i = 1, \ldots, d$ recorded at a site assumed to follow a reverse Weibull distribution $\tilde{F}_i$, where n_i is the number of zero wind speeds in direction i observed in n extreme wind speed events. Our objective is to estimate the parameters (α_i, η_i, c_i) of $\tilde{F}_i$. The method of moments, the method of maximum likelihood, the method of probability-weighted moments, and other methods can be used to estimate the parameters of this distribution ([4], Chapter 22). Extensive numerical studies suggest that the method of moments delivers satisfactory estimators for the unknown parameters of $\tilde{F}_i$, in contrast to, for example, the maximum likelihood method that can produce unstable estimators [6]. These features of the method of moments and its simplicity are our reasons for using it to estimate the parameters (α_i, η_i, c_i) of the distributions $\{\tilde{F}_i\}$. The following 3 step algorithm can be used to estimate the parameters (α_i, η_i, c_i).

Step 1. Construct the record $(y_1 = -v_{i,1}, y_2 = -v_{i,2}, \ldots, y_{n-n_i} = -v_{i,n-n_i})$. Since $(v_{i,1}, v_{i,2}, \ldots, v_{i,n-n_i})$ are independent samples of a reverse Weibull distribution $\tilde{F}_i$ with parameters (α_i, η_i, c_i), then $(y_1, \ldots, y_{n-n_i})$ are independent samples of a Weibull distribution with parameters $(\alpha = \alpha_i, \xi = -\eta_i, c = c_i)$.

Step 2. Calculate estimates $\hat{\mu}_y$, $\hat{\sigma}_y^2$, and $\hat{\gamma}_{y,3}$ for the mean μ_y, variance σ_y^2, and skewness coefficient $\gamma_{y,3}$ from the sample $(y_1, \ldots, y_{n-n_i})$. For example, $\hat{\mu}_y = \sum_{i=1}^{n-n_i} y_i/(n-n_i)$, $\hat{m}_q = \sum_{i=1}^{n-n_i} (y_i - \hat{\mu}_y)^q/(n-n_i)$ are estimates of the central moments of order $q \geq 2$, $\hat{\sigma}_y^2 = \hat{m}_2$, and $\hat{\gamma}_q = \hat{m}_q/\hat{\sigma}_y^{q/2}$ for $q = 3, 4$.

Step 3. Estimates the parameters (α, ξ, c) from Eq. 21. First, find an estimate $\hat{c}$ for c from the last equality in Eq. 21 with $\hat{\gamma}_{y,3}$ in place of $\gamma_{y,3}$. This nonlinear equation needs to be solved by iterations. Second, find an estimate $\hat{\alpha}$ for α from the second equality in Eq. 21 with (σ_y^2, c) replaced by $(\hat{\sigma}_y^2, \hat{c})$. Third, find an estimate $\hat{\xi}$ for ξ from the first equality in Eq. 21 with (μ_y, α, c) replaced by $(\hat{\mu}_y, \hat{\alpha}, \hat{c})$. If $\hat{\xi} > \min_{1 \leq j \leq n-n_i}\{y_j\}$, then set $\hat{\xi} = \min_{1 \leq j \leq n-n_j}\{y_j\}$ and calculate $(\hat{\alpha}, \hat{c})$ from the first equalities in Eq. 21 with $(\hat{\mu}_y, \hat{\sigma}_y^2)$ in place of (μ_y, σ_y^2). Set $(\alpha_i = \hat{\alpha}, \eta_i = -\hat{\xi}, c_i = \hat{c})$. Repeat these steps for all directions $i = 1, \ldots, d$ to determine completely the parameters of the marginal distributions of the wind speed vector $\boldsymbol{V}$.

The last ingredient of the probability law of $\boldsymbol{V}$ under the translation model in Eq. 1 is the covariance matrix $\boldsymbol{\rho}$ of $\boldsymbol{G}$. The covariance matrix $\boldsymbol{\rho}$ can be estimated from the Gaussian image of the available directional wind speed data defined by the mapping $G_i = \Phi^{-1}(F_i(V_i))$, $i = 1, \ldots, d$, in Eq. 1. A direct use of this mapping causes numerical problems since the distributions $\{F_i(x)\}$ have discontinuities at $x = 0$. We eliminate this computational problem by replacing zero readings in the data set with random noise of small intensity. Specifically, we replace zero wind speeds with independent samples of a uniform random variable with support $(0, \varepsilon)$, where $\varepsilon > 0$ is a parameter smaller than all non-zero readings. Accordingly, the mapping $V_i \mapsto G_i = \Phi^{-1}(F_i(V_i))$ is replaced by $V_i \mapsto G_i^* = \Phi^{-1}(F_i^*(V_i^*))$, where

$$F_i^*(x) = q_i \left[\frac{x}{\varepsilon} 1(0 \leq x \leq \varepsilon) + 1(x > \varepsilon) \right] + (1 - q_i) \tilde{F}_i(x), \quad i = 1, \ldots, d, \qquad (22)$$

$\tilde{F}_i$ is as in Eq. 1, $\boldsymbol{V}^*$ is the modified wind speed vector, and $\boldsymbol{G}^*$ is the Gaussian image of $\boldsymbol{V}^*$. The covariance matrix $\boldsymbol{\rho}^*$ of $\boldsymbol{G}^*$ can be estimated from the Gaussian image of the modified wind speed data by using standard statistical tools.

There are two sources of uncertainty in the estimated probability law of $\boldsymbol{V}$, the marginal distributions $\{F_i\}$ and the covariance matrix of the Gaussian image of $\boldsymbol{V}$. Since we postulate the functional forms of $\{F_i\}$, the uncertainty in these distributions is limited to the uncertainty in their estimated parameters. The construction of the Gaussian image of the wind speed data involves modifications of the available wind record, so that there are modeling errors. However, these errors are not relevant for design since they relate to very small wind speeds. The modified and original wind data coincide for wind speeds relevant for design. We use parametric bootstrapping to quantify approximately effects of the uncertainty in the parameters of the probability law $\boldsymbol{V}$ on wind speeds and structural demands of various MRI's.

4.2 MONTE CARLO ALGORITHM

Once the probability law of the proposed directional wind speed model has been calibrated to a wind data recorded at a site, we can apply the following algorithm to generate directional wind speed records of any length that are consistent the date set. Suppose our objective is to generate directional wind speed at a site under consideration over τ years. If λ is the average number of extreme wind events in a year, the average number of extreme wind events in τ years is $n_\tau = [\lambda \tau]$ where $[a]$ denotes the largest integer smaller than a. Accordingly, the extreme wind environment of the site is described by n_τ events.

The following three step algorithm can be used to generate extreme wind environments at a site that are consistent with the available site data.

- *Step 1.* Generate samples $\boldsymbol{g}^* = (g_1^*, \ldots, g_d^*)$ of a d-dimensional standard Gaussian vector $\boldsymbol{G}^*$ with covariance matrix $\boldsymbol{\rho}^*$. Various methods can be used to generate independent samples of $\boldsymbol{G}^*$ ([2], Section 5.2.1).

- *Step 2.* Calculate the image $\boldsymbol{v}^* = (v_1^*, \ldots, v_d^*)$ of $\boldsymbol{g}^*$ defined by the mapping

$$v_i^* = \left(F_i^*\right)^{-1}\left(\Phi(g_i^*)\right), \quad i = 1, \ldots, d. \tag{23}$$

- *Step 3.* Repeat Steps 1 and 2 n_τ times to generate a time series of directional extreme wind speeds over τ years. The time arrivals of these extreme wind events are samples of a Poisson process N of intensity λ.

Chapter 5
WIND MODEL FOR THUNDERSTORM AND SYNOPTIC WINDS

===

Available data for thunderstorm and synoptic winds can be insufficient to estimate the dependence between directional wind speeds and even the three parameters of the reverse Weibull distribution we used for hurricane winds. It is assumed that directional wind speeds are independent and have distributions defined by Eq. 17, where $\tilde{F}_i$ are the Extreme Type I distributions

$$\tilde{F}_i(x) = \exp\left\{ -\exp\left[-\alpha_i\left(x - u_i\right)\right]\right\}, \quad x \in \mathbb{R}, \quad i = 1, \ldots, d. \tag{24}$$

The means $\{\mu_i\}$ and standard deviations $\{\sigma_i\}$ of directional wind speeds are related to the defining parameters (α_i, u_i) by the relationships

$$\alpha_i = \frac{\pi}{\sqrt{6}\,\sigma_i}$$
$$u_i = \mu_i - \frac{0.577216}{\alpha_i}. \tag{25}$$

The defining parameters of the probabilistic model for thunderstorm and synoptic winds are the mean arrival rate λ of these wind events and the parameters $\{(\alpha_i, u_i), i = 1, \ldots, d\}$. The mean arrival rate is estimated following the procedure for hurricanes. The relationships in Eq. 25 can be used to find $\{(\alpha_i, u_i), i = 1, \ldots, d\}$ from estimates of directional wind speed means and standard deviations.

For the data sets used in the report, most wind directions have only zero readings, so that the resulting estimates of q_i and μ_i are zero. In the absence of additional information, the user may choose to assume that these estimates are caused by insufficient data, and that q_i and μ_i on these directions should be similar to neighboring non-zero estimates. Accordingly, regressions are developed for $\{\mu_i\}$ and $\{q_i\}$. For the directions with a sufficiently string of non-zero wind speeds, the estimates of q_i and μ_i calculated from data are retained. For the other directions q_i and μ_i are those given by regressions. The coefficient of variation corresponding to the direction with the largest number of non-zero readings is used in all directions. With these assumptions the model for directional wind speeds in thunderstorm and synoptic events is completely defined. We note that these considerations are heuristic. Alternative procedures can be used to assign non-zero values to q_i and μ_i in directions with only zero readings.

Samples of directional wind speeds for thunderstorm and synoptic events can then be obtained by elementary calculations. First, we generate d independent random numbers $(r_1, \ldots, r_d)$ uniformly distributed in $(0, 1)$. If $r_i \leq q_i$, the wind speed in direction i is set zero. Otherwise, the wind speed in this direction is (Eq. 24)

$$x = u_i - \ln\left(-\ln(r)\right)/\alpha_i \tag{26}$$

where r is a random number uniformly distributed in $(0, 1)$.

Chapter 6
NUMERICAL ILLUSTRATION FOR HURRICANES

===

The directional extreme wind speed series at Miami, milepost 1450, consists of $n = 999$ events with wind speeds measured in $d = 16$ directions. This data set is used to calibrate the proposed model for directional extreme wind speeds, calculate wind speeds over a broad range of MRIs irrespective of direction, and assess the uncertainty in the generated wind speeds by parametric bootstrapping. The defining parameters of the directional extreme winds are the mean arrival rate λ of events per year, the probabilities $\{q_i = P(V_i = 0)\}$ of directional wind speeds being zero, the parameters of the distributions $\{\tilde{F}_i\}$ of the non-zero values of directional wind speeds $\{V_i\}$, and the covariance matrix ρ^* of the Gaussian image G^* of the modified wind speed vector V^*.

6.1   MODEL CALIBRATION

The mean arrival rate of hurricane at this milepost is $\lambda = 0.56$ events per year, so that $n = 999$ hurricanes span on average over a period of $999/0.56=1784$ years. The estimates of the probabilities $q_i = P(V_i = 0)$, $i = 1, \ldots, 16$, are in the range $[0.4140, 0.9450]$, so that the number of non-zero values of directional wind speeds $\{V_i\}$ in the data set is in the range $[55, 585]$. Accordingly, the estimated parameters of the distributions $\{\tilde{F}_i\}$ of the non-zero values of $\{V_i\}$ will have different accuracies.

The Gaussian image G^* of the modified wind speed data set at Miami has been obtained from the mapping defined by Eq. 22 with $\varepsilon = 0.1$, a value much smaller than values of wind speeds relevant for design.

6.2   MONTE CARLO SIMULATION

The Monte Carlo simulation algorithm outlined in a previous section has been applied to generate directional wind speed sequences with independent and dependent coordinates. Both the independent and dependent wind speed vectors have the same marginal distributions. The dependence between directional wind speeds correspond to that between the coordinates of the Gaussian vector G^*.

Results are reported for the series of the largest wind speeds $\{V_{k,\max}, k = 1, 2, \ldots\}$ irrespective of direction. Similar results can be obtained for the sequence of structural demands $\{D_s^{(k)}, k = 1, 2, \ldots\}$ at an arbitrary structural point s caused by a series of directional extreme winds $\{V^{(k)}, k = 1, 2, \ldots\}$.

Figure 3 shows with heavy solid and dotted lines wind speeds v_R of MRIs up to 1,800 years in Miami obtained by the proposed Monte Carlo algorithm with dependent and independent directional wind speeds. As previously stated, the calculations are based on Eq. 12. The assumption of independence between directional wind speeds is conservative in agreement with results in Fig. 1. The thin solid line in the figure gives estimates of v_R obtained directly from data. Both model-based estimates of v_R exceed those obtained directly from

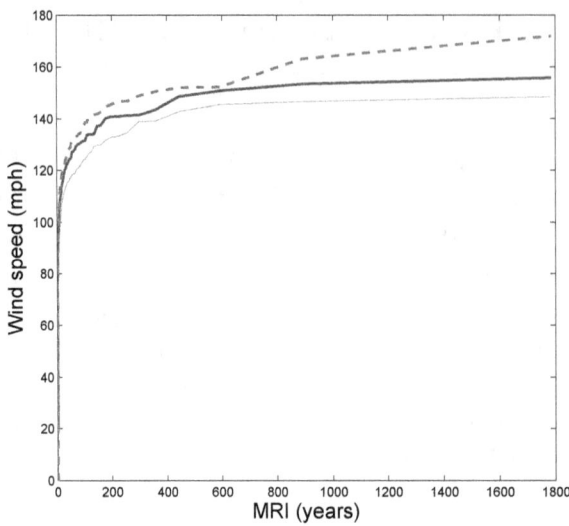

Figure 3: Data and model-based wind speeds in Miami for MRIs up to 1,800 years

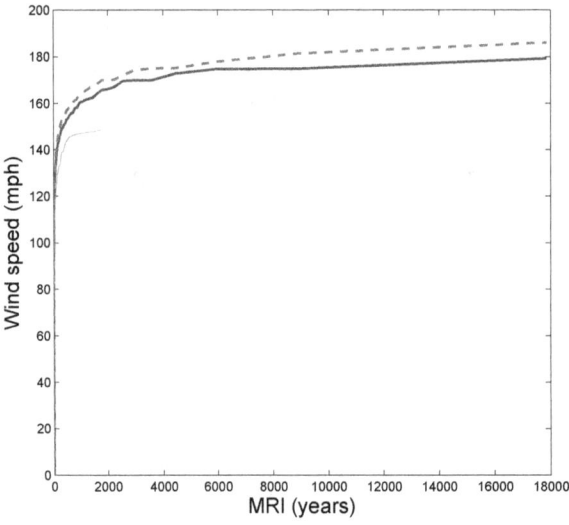

Figure 4: Data and model-based wind speeds in Miami for MRIs up to 18,000 years

data. That model-based estimates of v_R are larger than those obtained directly from data may be explained by the way in which the dependence between directional wind speeds is represented. The dependence between directional wind speeds is ignored in the independent model and is likely to underestimate the actual dependence in the dependent model since we replace all zero readings in the data set with independent random variables. Figure 4 shows similar plots as in Fig. 4 for larger MRIs. Since available data is limited, wind speeds of large MRIs cannot be obtained directly from data, as illustrated by the thin solid line in this figure which cannot be continued beyond approximately 1,800 years. On the other hand, the

18

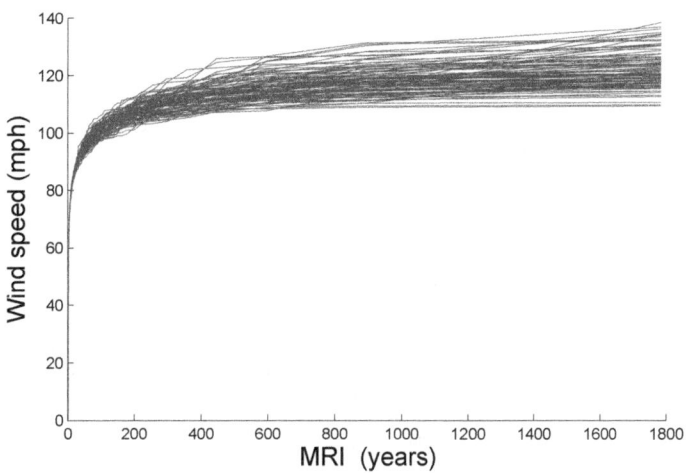

Figure 5: One hundred replicates of model-based wind speeds in Miami for MRIs up to 1,800 years

proposed Monte Carlo algorithm can be used to generate wind speeds of any MRI.

6.3   UNCERTAINTY IN MONTE CARLO ESTIMATES

We assess the statistical uncertainty in our estimates of the wind speeds v_R by bootstrapping. Classical bootstrapping using empirical distributions to generate replicates of wind speed series is inadequate since v_R is likely to be outside the range of data and all replicates would take values in the range defined by data. To overcome this limitation, we use parametric bootstrapping, that is, replicates are generated from probabilistic models for directional wind speeds calibrated to data rather than empirical distributions given by data. Accordingly, we use the Monte Carlo algorithm outlined in a previous section to generate replicates of the directional extreme wind speeds at a site. Figure 5 shows 100 replicates of the heavy solid line in Fig. 3, that is, wind speeds v_R with return periods up to 1,800 years. A histogram of the v_R for an MRI R of approximately 1,786 years is shown in Fig. 6, and is based on 100 replicates wind speeds v_R produced by the proposed Monte Carlo algorithm. The range of v_R in these 100 replicates is approximately $[110, 138]$ mph. Larger number of replicates can be used to obtain ranges including v_R with specified probabilities.

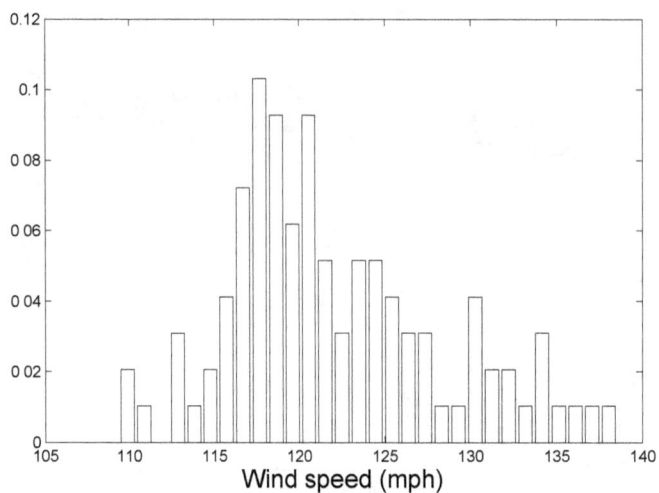

Figure 6: Histogram of v_R for MRI= $1,786$ years

Chapter 7
CONCLUSIONS

===

Directional wind speeds for periods much longer than commonly available records are needed to estimate wind effects in a variety of structural systems including tall buildings. Since wind records of the requisite size are not available, we propose to generate synthetic directional wind speed records of the requisite length from probabilistic models calibrated to data.

It has been shown that probabilistic models can be constructed for directional wind speeds observed in hurricanes, thunderstorm, and synoptic winds, and that these models can be calibrated to data. Since available records for thunderstorm and synoptic winds can be relatively short and have numerous zero-readings, a simplified probabilistic model has been used to describe their directional wind speeds. The model is based on the assumption that the non-zero directional wind speeds are independent Extreme Type I variables. On the other hand, the probabilistic model for directional wind speeds in hurricanes (or thunderstorms and synoptic winds for which sufficient observations are available) accounts for the dependence between directional wind speeds.

The probabilistic models for directional wind speeds in hurricanes, thunderstorm, and synoptic winds have been used to develop MATLAB functions for (1) generating directional wind speeds in extreme wind events spanning arbitrary time intervals and (2) assessing the uncertainty in the generated wind speeds by bootstrapping. Appendices A and B outline the use of these MATLAB codes and their main features.

Appendix A
GENERATION OF DIRECTIONAL WIND SPEEDS
==

A MATLAB function **dir_hurricane** has been developed to generate wind speeds in a specified number **ns** of extreme wind events that are consistent with data for hurricane, thunderstorm, and synoptic winds. The input and output variables for **dir_hurricane** are defined in this function, and their definition is not repeated. We only mentioned that the generated wind speeds are collected in (ns, d)-matrices called **w_speed_MC** and **w_speed_IM_MC**.

- For hurricane winds, the wind speeds in **w_speed_MC** and **w_speed_IM_MC** correspond to models accounting for and disregarding the dependence between directional wind speeds, respectively.

- For thunderstorm and synoptic winds, only independent directional wind speeds are generated because of the limited number of data. The output matrices **w_speed_MC** and **w_speed_IM_MC** are identical for these winds.

Directional wind speed data are loaded via a dialog allowing to select wind types and sites. The steps for this selection are explained in the code.

A.1   MODEL FOR HURRICANE WINDS

It is assumed that directional wind speeds in hurricane winds follow reverse Weibull distributions. Translation random vectors are used to capture the dependence between directional wind speeds. The calibration of this model in the MATLAB function **dir_hurricane** follows the steps outlined in Sec. 4.1. The sample generation follows the algorithm in Sec. 4.2.

The d-dimensional output vector **qv** gives estimates of the directional probabilities $\{q_i\}$ in Eq. 17. The d-dimensional output vectors **alw, cw**, and **xiw** give estimates of the parameters $\{\alpha_i\}$, $\{\eta_i\}$, and $\{c_i\}$ of the directional reverse Weibull distributions defined in the context of Eqs. 18 and 19. The (d, d) matrix **covgxnew_est** is an estimate of the covariance matric $\boldsymbol{\rho}^*$ of the standard Gaussian vector $\boldsymbol{G}^*$ defined in Sec. 4.1. The (ns, d)-output matrices **w_speed_MC** and **w_speed_IM_MC** have already been defined.

A.2   MODEL FOR THUNDERSTORM AND SYNOPTIC WINDS

It is assumed that directional wind speeds in thunderstorm and synoptic follow a distribution of the type in Eq. 17 but they are independent and $\tilde{F}_i$ are the Extreme Type 1 distribution defined by Eqs. 24 and 25.

The input parameters **cmin, cmax, nc, epsi, epsi_1** are not needed when dealing with thunderstorm and synoptic winds; any values can be used for these parameters. The d-dimensional output vector **qv** gives estimates of the directional probabilities $\{q_i\}$. The d-dimensional output vectors **alw** and **xiw** give estimates of the parameters $\{\alpha_i\}$ and $\{u_i\}$ of the distributions $\{\tilde{F}_i\}$ in Eq. 24. The output parameters **cw** and **covgxnew_est** have no meaning. As previously state, the (ns, d)-output matrices **w_speed_MC** and **w_speed_IM_MC** with generated directional wind speed data coincide.

Appendix B
PARAMETRIC BOOTSTRAPPING

==

Two MATLAB functions **bootstrap_par** and **bootstrap_par_ts** are used to assess our confidence in hurricane and in thunderstorm/synoptic wind speeds generated by **dir_hurricane**. The input and output for **bootstrap_par** and **bootstrap_par_ts** are defined in these functions, and their definitions are not repeated. We only mention that **w_speed_boot** is a three-dimensional array of dimension (n, d, nb), where $n = $ the number of extreme wind events, $d = $ the number of wind events, and $nb = $ the number of replicates.

To run **bootstrap_par** and **bootstrap_par_ts**, the output of **dir_hurricane** needs to be stored in the file **boot** for hurricane winds and the file **boot_ts** for thunderstorm or synoptic winds. The **boot** file includes the vectors **qv, alw, cw, xiw** and the matrix **covgxnew_est**. The **boot_ts** file includes the vectors **qv, alw, xiw**.

References

[1] M. Grigoriu. *Applied Non-Gaussian Processes: Examples, Theory, Simulation, Linear Random Vibration, and MATLAB Solutions.* Prentice Hall, Englewoods Cliffs, NJ, 1995.

[2] M. Grigoriu. *Stochastic Calculus. Applications in Science and Engineering.* Birkhäuser, Boston, 2002.

[3] N. Johnson and S. Kotz. *Distributions in Statistics: Continuous Univariate Distributions-1.* Houghton Mifflin Company, Boston, MA., 1970.

[4] N. L. Johnson, S. Kotz, and N. Balakrishnan. *Continuous Univariate Distributions*, volume 2. John Wiley & Sons, Inc., New York, 1994. Second Edition.

[5] M. R. Leadbetter, G. Lindgren, and H. Rootzén. *Extremes and Related Properties of Random Sequences and Processes.* Springer-Verlag, New York, 1983.

[6] E. S. Martins and J. R. Stedinger. Generalized maximum-likelihood generalized exteme-value quantile estimators for hydrological data. *Water Resources Research*, 36(3):737–744, 2000.

[7] E. Simiu and T. Miyata. *Design of Buildings and Bridges for Wind.* John Wiley & Sons, Inc., New York.

[8] E. Simiu and R. Scanlan. *Wind Effects on Structures: Fundamentals and Applications to Design.* John Wiley & Sons, New York, NY, 1996.